BUILDING CALIFORNIA

Technology and the Landscape

Michael R. Corbett, author

Waverly Lowell and Emily Wolff, curators

California Historical Society, San Francisco
National Archives and Records Administration, Pacific Region
Documents Collection, College of Environmental Design, University of California, Berkeley
The Foundation for San Francisco's Architectural Heritage

1998

Library of Congress Catalog Card Number 98-60287
ISBN 0-9651144-7-3

Published on behalf of the California Historical Society by William Stout Publishers, 501 Greenwich Street, San Francisco, California 94133

Printed in Salt Lake City, Utah, USA

CONTENTS

This volume serves as the catalogue of an exhibition by the same name, held at the California Historical Society, San Francisco, from May 7, 1998 to August 15, 1998. The exhibition and catalogue were made possible with funds provided by:

Bechtel Group, Inc., "Building a Century: 1898-1998"
The Bernard Osher Foundation
The Foundation for San Francisco's Architectural Heritage
Grants for the Arts/San Francisco Hotel Tax Fund

and also by the individual contributions of the following:

Architectural Resources Group
Charles Edward Bassett
Carey & Co. Inc. Architecture
Joseph Esherick
Marty Gordon
Harry E. Miller, in memory of Charles and Mary Catherine Page
Oliver & Company
C. David Robinson Architects
Page & Turnbull
Rosekrans & Associates, Architects
James Swinerton

The exhibition and catalogue were organized with the assistance of an Advisory Committee composed of the following members:

David Bahlman, The Foundation for San Francisco's Architectural Heritage
Charles Edward Bassett
Alice Carey, Carey & Co. Inc. Architecture
Joseph Esherick
Stephen Farneth, Bruce Judd, Bridget Maley, Architectural Resources Group
Marty Gordon
Clifford Graves
Daniel Gregory, *Sunset Magazine*
Richard Longstreth
Ian Mackinlay, Ian Mackinlay Architecture
Harry E. Miller
Steven H. Oliver, Oliver & Company
Charles Hall Page, J. Gordon Turnbull, Page & Turnbull
Mrs. G. Bland Platt
C. David Robinson
Adolph Rosekrans
Gregory Ryken
James Swinerton

Director's Foreword

Where we live and work are timeless topics of discussion, central to us all. Every day, we look at the buildings that surround us and see "what" is there, making mental notes as to type and style. We may think about what is pleasing, and what is not, about a particular structure, but rarely do we ask "why" a structure was built the way we see it.

"Why" is such a magical word because it opens up questions. *Building California: Technology and the Landscape*, goes beyond "what" we see to take up the question of "why" the architecture of California is built the way it is. By analyzing the environmental and economic factors that affect building, the availability of materials, the organization of the labor force, and the impact of natural disasters, the essay contained in this catalogue provides us with a new perspective for looking at buildings. With its useful illustrations and extensive bibliography, this catalogue will make a valuable addition to every library, whether personal or professional.

The California Historical Society is honored to join with the Foundation for San Francisco's Architectural Heritage; the National Archives and Records Administration, Pacific Region; and the Documents Collection, College of Environmental Design, University of California, Berkeley, in organizing this catalogue, the companion piece to an exhibition by the same name. Architectural historian Michael Corbett provided the initial idea for the exhibition and wrote the catalogue essay; his clear and insightful treatment of this vast subject was key to the success of this enterprise. Of equal value was the work of the curators, Waverly Lowell, former Director of the National Archives and now Director of the Documents Collection at the College of Environmental Design, and Emily Wolff, Manager of the Society's Photography Collection. Their professionalism, hard work, and humor made the exhibition and catalogue possible.

I am indebted to many, including the staff and Board of Trustees of the California Historical Society and the Advisory Committee, whose members gave encouragement and guidance to the project from the very beginning. Together, we hope that *Building California: Technology and the Landscape* will reveal to the reader new dimensions of the world around us.

Michael McCone, Executive Director
California Historical Society

INTRODUCTION

This essay presents a brief overview of the history of structures and materials of California building, from pre-colonial times to the close of World War II. This is a history of materials—wood, brick, iron, steel, and concrete—and the technologies employed to build with them. It is also a history of the relationship between building technologies and the natural and cultural landscape. Here, 'landscape' is understood in the terms suggested by J. B. Jackson, as the visible record of human beings living on the land, reflecting their needs, attitudes, and values.[1] Building technologies and the landscape are related, then, both in their physical features and in the intangible influences that shape them. This essay attempts to address the whole range of connections between building technologies and the world around them.

Much of what has been written about Northern California architecture in the past has focused on the regional character of its architectural styles—the distinctive exuberance of the Victorian house, for instance, or the innovations of the Bay Area Tradition. This attention to style and to regionalism has distorted the larger picture of California architecture by emphasizing superficial differences, while neglecting deeper ones. The view of structures and materials presented in this essay suggests that, in most ways, California buildings are indistinguishable from buildings built in other parts of the country. At the same time, this essay demonstrates that there are differences that are based on a response to natural conditions and phenomena, such as earthquakes, fires, landslides, and floods, and that these differences represent a different kind of regionalism.

The goal of this essay is to identify the issues central to the history of structures and materials in Northern California and to provide a context for understanding individual buildings. The examples cited in the text illustrate aspects of the subject and do not constitute a comprehensive history. Although these examples focus primarily on the architecture and building conditions of San Francisco, the ideas presented apply to Northern California generally. By examining both the construction of individual buildings and the development of cities, towns, and the countryside, this essay will help to form a more complete picture of California architecture.

Native Californians constructed buildings using local materials and tools. These Northern Pomo dwellings were built of lumber split from fallen trees.

NATIVE CALIFORNIA, SPAIN AND MEXICO

The area now known as California was occupied by native peoples for at least ten thousand years before the arrival of colonizers, sent from Mexico by the Spanish government in 1769.[2] Although the authority for these colonizers gave way to that of Mexico in 1821, California was culturally similar under both governments. The Spanish-Mexican period in California ended in the late 1840s with the military defeat of Mexico by the United States, followed by the Gold Rush. During the period between 1769 and 1849, both human society and building traditions underwent enormous change. The change from Native American to Spanish-Mexican society almost destroyed the Native American world in the coastal area occupied by colonizers. The transition from Mexican to American society corresponded with the rapid transition from pre-industrial to industrialized society.

At the time of Columbus, California was the most densely populated part of what would become the United States, with more than five hundred different groups of people speaking 120 languages. The building traditions of these diverse peoples produced many different types of building. This variety was achieved even though building efforts were limited by the availability of materials and tools.

The structures and materials used in Native American buildings varied with the availability of usable tools. They also differed by locality, and within locality, by season, depending on climate, precipitation, soil, and vegetation. Northern California possesses unusually diverse natural landscapes: the foggy coastal range along the Pacific Ocean, with its redwood forests; the alluvial plains around San Francisco and Monterey bays, with their stream courses, marshes, and hills; the great Central Valley, with its tule marshes and vast seasonal lakes; the Sierra Nevada-Cascade mountain range; the dry east side of the Sierra Nevada; and the rugged, forested Klamath mountains. Many native peoples moved during the seasons among these different types of landscapes, building structures appropriate to each place and each season.

Around San Francisco Bay and the rivers and marshes of the Central Valley, the Ohlone and other groups built round dwellings on frames made of willow poles that were thatched with bunches of dried tules. In the redwood forests along the northern coast, the Hoopa built sturdy, rectangular houses

Spanish builders imported masonry construction technologies from Mexico and Spain. Most buildings of the period were constructed of adobe blocks, made by hand from local sources of clay and straw.

of split wood planks or slabs of bark separated from fallen logs with wood levers, elk antlers, or sharpened pieces of stone. In the coastal mountains north of San Francisco, the Pomo built large, round, ceremonial lodges out of poles covered with earth.

The built landscape of pre-colonial California consisted of villages and small camps. Native California villages included dwellings, granaries, cisterns, ceremonial lodges, sweat houses, and menstrual huts. Some peoples, like the Yurok, built large, permanent villages; some Yurok houses are believed to have been occupied for more than two thousand years. The villages of most peoples were small and houses were built to survive only a short time—from a single season to a few years. The brush huts of hunting groups may have been built to last only a few weeks. Because they were built of materials found within their surroundings, buildings and villages shared the muted colors of the natural landscape. Once abandoned, most sites quickly reverted to their natural condition.

When the Spaniards arrived in Northern California in the 1770s, they brought with them building technologies from Spain and Mexico.[3] Like the native peoples of the region, they employed local materials and relied on human labor. The Spaniards, however, built larger, permanent structures with more complex spatial configurations. Building practices involved the use of specialized skills, a large labor force, and a hierarchical division of labor. Buildings were designed and built under the guild system, with master masons or master carpenters at the top. The work of these highly-skilled artsans was supported by assistants, called journeymen, and by specialists in other areas, such as blacksmiths and painters. Spanish and Indian workers provided physical labor.

The most common type of construction under the Spaniards used a masonry technology indigenous to Spain and Mexico. Adobes—blocks made of clay and straw pressed into wooden molds and dried in the sun—were laid up with

Skilled artisans trained in Mexico supervised an organized labor force. Due to a shortage of masons, San Carlos Borromeo in Carmel was one of the few buildings of the period to be constructed in stone.

mud mortar in masonry walls atop stone foundations. Flat roofs were made of kiln-fired bricks called *ladrillos*, placed on wood beams covered with straw and dirt. Alternatively, pitched roofs were constructed of wood and covered with thatch or tiles. Tiles were formed of clay, sand, and straw, that was pressed into molds and fired in kilns. Mortar for stone masonry was made from lime reduced in kilns from limestone or seashells. Plaster used to waterproof flat roofs and walls was also made of lime. This lime-based coating was white in color.

Two of the most substantial buildings in Northern California—the Royal Presidio Chapel in Monterey and the mission San Carlos Borromeo in Carmel—were constructed of stone. During the construction of each, sandstone quarried nearby was hauled by ox-drawn carts to the building site. The master mason of both buildings, Manuel Esteban Ruiz, assisted by journeymen and Indians who were masons-in-training, cut stones with iron tools. These stones were laid up with mortar made from shells to form walls and then vaulted, originally, to form the roofs. The roofs were clad in tiles, the walls plastered, the interior walls frescoed, and the buildings embellished with carved wood doors, altars, and furnishings.

The use of local materials in the Spanish-Mexican period produced buildings and built landscapes that were frequently similar in color to their natural surroundings. In contrast to the buildings and settlements of Native Americans, however, those of the Spanish and Mexicans were intended to be permanent and generally lasted at least a few years, rather than a single season. They were larger and formed conspicuous features of the landscape. In many cases, they were made even more conspicuous by the use of whitewash.

Many of the buildings visible in the earliest photographs of San Francisco were prefabricated structures. These structures were brought to California by ship to supply the needs of the Gold Rush population.

THE GOLD RUSH

The Gold Rush brought tens of thousands of people to California in the course of a few months. This sudden influx of people affected all aspects of life, but in particular, it overwhelmed the capacity of the region to satisfy the vast need for houses and other types of buildings.[4] The conditions created by the Gold Rush inaugurated a radically new era of building in California.

During the decade prior to the Gold Rush, California had begun the transformation from a Native American and Mexican society to an American society. Following the secularization of the Spanish missions, an increasingly large number of outsiders, especially Americans, entered California. These outsiders brought with them new skills and started up new economic enterprises. They engaged in logging and built sawmills. Sawn lumber became one of California's principal commodities during the 1840s; it was produced by hired labor and sold as far away as Hawaii. Immigrants such as Thomas D. Larkin influenced design and building practices by introducing wood construction techniques from New England.[5] But the production of lumber and Larkin's new approaches to building were wholly inadequate in satisfying the new and sudden demand for buildings. The changes that were set in motion by the Gold Rush were far greater than any that had come before.

The Gold Rush boosted California's nascent commercial economy. In sharp contrast to the largely pre-industrial methods and materials used by the Indians, Spaniards, Mexicans, and Americans before them, the builders of the Gold Rush period began to rely on industrialized building methods and on materials produced by industrial means. For the first time in California's history, building materials were not produced locally, but came from around the world. Prefabricated buildings and materials were brought to California aboard ships. Although few buildings from this period survive, the efforts to obtain buildings and materials reflect a development in both commerce and industry that would have lasting impact on the built landscape of California.

In the chaos of the first few months of 1849, the people of the Gold Rush resorted to temporary or makeshift solutions to satisfy their immediate need for shelter. Among these makeshift solutions were tents, lean-tos, and shanties. In addition, abandoned ships were inhabited with little or no physical alteration to their structure, simply as they floated or lay aground in the tidal flats, accessible only by rowboat at high tide. At least a few ships were incorporated as the first stories of taller buildings, with additional stories constructed atop their decks. Other ships were cannibalized for parts, their wooden decks, bunks, and hulls used as framing, siding, roofing, and floorboards in new construction. Discarded sails served as interior partitions and as window and door coverings within these buildings, and as covers for tents and lean-tos as well.

More important in establishing a long-term capacity for building in California was the creation of a market in building materials. A significant manifestation of this market was the large-scale importation of prefabricated structures to California. In the long history of building prefabrication, the first few months of the California Gold Rush stand out as an unusually profitable period; no doubt the makers of these buildings hoped that the market for their product would continue to grow.[6] Although some prefabricated buildings were sent to California on order, many were sent on a speculative basis.

Prefabricated buildings had been shipped to Pacific destinations early in the nineteenth century, but the Gold Rush created a boom. Prefabricated buildings were shipped from manufacturers around the world, including the East Coast of the United States, several European countries, New Zealand, Australia, and China. These buildings came by sea, either across the Atlantic Ocean and around Cape Horn, or across the Pacific. Sometimes they were shipped to Central America and carried overland, across the Isthmus of Panama, and then reloaded on ships bound for California. The majority of prefabricated buildings ended up in San Francisco: the earliest daguerreotypes and photographs of that city illustrate what appears to be an abundance of prefabricated buildings. Some prefabricated structures were also erected in San Jose, Marysville, Stockton, Sacramento, and in the mining regions scattered throughout California.

Of the many types of prefabricated buildings that were constructed, houses were the most common. Among the best-known surviving examples are two of three houses ordered by General Mariano Vallejo; his own house, known as *Lachryma Montis*, in Sonoma, and the Frisbie-Walsh House in Benicia, both of 1849. Hotels, warehouses, stores, hospitals, and a bowling alley were also sent to California to be assembled from prefabricated parts; the largest of these may have been the three-and-one-half story, 100-room hotel, Astor House, erected in San Francisco.

There was as much variety in the structure and appearance of prefabricated buildings as in manufacturers and places of manufacture. Buildings were constructed primarily of two materials—wood and metal. Many of the wooden buildings appear to have been assembled using brace-frame construction. Like the traditional New England house, these prefabricated buildings were constructed using a frame of heavy timbers, held together by means of mortise and tenon joints. Other wood-frame buildings relied upon a simplified version of the labor-intensive, brace-frame system; their frames were assembled, wholly or in part, with milled lumber and nails. Some buildings were constructed in the Gothic Revival style, with pointed-arch windows and decorative barge

Prefabricated buildings came in many different sizes and designs from manufacturers in at least a half-dozen countries. Wood and corrugated iron buildings were shipped to San Francisco and to many inland locations.

boards; others may have been built in the Greek or Classical Revival style. Most, however, were unornamented, although endlessly varied in appearance due to their size, proportions, roof pitch, and fenestration.

Greater conceptual variety of structure characterized the prefabricated iron buildings that reached California. One type of house, manufactured in New York, was constructed of parts designed to slide together by means of grooves. Another type, made in Philadelphia, consisted of cast iron columns covered with an exterior sheathing of riveted metal plates. A common type, easily recognized in photographs, had walls and a semicircular roof made of corrugated, galvanized iron. Among the claims made by the advertisers of these products was that these buildings were fireproof and could be taken down and reassembled in new locations.

With few exceptions, the buildings of the early Gold Rush—whether intended as temporary or permanent—did not survive. As discussed in the next section of this essay, most of these buildings were destroyed by fire. But although the physical remnants of the period may have been lost, the materials and methods of production that were established during this period continued, as did the international economy that supported their development. Virtually everything built since the Gold Rush relies upon the industrial technologies introduced at this time.

PLACERVILLE.

Lith. & Published by Quirot & Co. corner Montgomery & California Sts. S. Francisco.

The Gold Rush resulted in the rapid establishment of cities within a wide area of Northern California. The collective demand for building materials created new industries, beginning with lumbering.

INSTANT SETTLEMENT

Three principal types of places emerged in Northern California during the first few months of the Gold Rush: mining camps, such as Weaverville, Nevada City, Grass Valley, Auburn, and Placerville; bay and river ports such as San Francisco, the region's principal port; and inland distribution centers, including Sacramento, Stockton, and Marysville.[7] By 1854, the end of the placer mining period, more mining towns had been founded and a new type of settlement—the agricultural center—had come into being, especially in the flat lands around the Bay. In these places, substantial settlements were established in a matter of weeks or months, so quickly that they have been called "instant cities."[8] In order for this to have taken place, building materials and labor had to be made readily available, and buildings had to be built quickly.

In less than a year's time, a building industry sprang up locally to supply the needs of the booming population. The cities that began in 1849 with tents, shanties, and prefabricated buildings were rebuilt, primarily with buildings constructed of wood. Lumber was the first building material to be produced locally because wood was plentiful in the uncut forests around San Francisco Bay and along the coast. Large, permanent sawmills were established in Marin County, Oakland, Redwood City, and Eureka for the milling of redwood logs cut from the nearby hills. Small, portable sawmills were used in the mining camps and moved to fresh forests periodically, as local trees were cut down. Barges, loaded with lumber and docked at bay and river ports, were common features of the California landscape from the 1850s until the turn of the century.

The first new buildings made possible by the development of local industry may have been built using a variant of traditional brace-frame construction. In a traditional braced frame, columns and beams are made of timbers split from logs by hand, as in the James Johnston house in Half Moon Bay of 1853. This type of construction requires highly skilled workers; each piece of a building's frame is individually shaped for a specific location within the overall structure, and as a result, each building is unique. Loads are carried on heavy timber columns, typically six inches square. These columns are connected to slightly smaller, horizontal beams and diagonal braces by means of mortise and tenon joints. A brace-frame building is as strong as its frame.

The rapid construction of San Francisco in the second half of the nineteenth century depended on the use of balloon-frame construction and on innovations in the manufacture of lumber and building parts, such as windows and doors.

Once a frame is completed, it is enclosed with horizontal weather boards. Brace-frame construction, and simplified forms of brace-frame construction that substitute nails for mortise and tenon joints, continued to be used in California for barns and other agriculture buildings into the twentieth century. With mechanized sawmills, the preparation of structural timber columns and beams for use in brace-frame buildings was made easier and more efficient, and resulted in the rapid construction of many buildings.

Knowledge of a simpler, lighter form of construction spread to California at some point during the Gold Rush. Developed in the Midwest during the 1830s, this form of construction, called balloon-frame construction, uses lumber cut in standard dimensions and assembled with nails. Instead of widely spaced, heavy columns, a balloon frame consists of a series of two-by-four-inch studs, spaced sixteen inches apart. The studs rise from the sill, located at the foundation, to the plate, located at roof level, up to a height of perhaps three stories. This light frame is clad, either inside or outside, in a sheathing of diagonal boards. When the interior walls, typically of wood lath and plaster, and the exterior siding of horizontal boards are attached, balloon-frame construction is stronger than traditional brace-frame construction.

By 1900, a variant of the balloon frame, called the western platform frame, was in use in California. In platform-frame construction, the studs supporting the walls span only the height of a single floor, rather than the height of the entire building. This same type of construction is still used for wood houses in the 1990s.

The balloon frame eventually replaced brace-frame construction, although exactly when this took place is not known.[9] The rapid construction of rows of houses during the early 1870s by the Real Estate Associates in San Francisco

could only have taken place by employing balloon-frame construction.[10] Understanding of balloon-frame construction was spread, in part, by pattern books such as Woodward's *Country Homes*, first published in 1865. The first popular pattern books produced in California—John Cotter Pelton's *Cheap Dwellings* of 1882 and Samuel and Joseph C. Newsom's *Picturesque California Homes* of 1884—provided information on construction. Pelton illustrated plans for barns using brace-frame construction and for houses using both balloon-frame and single-wall construction. Single-wall construction, often signaled by board-and-batten siding, consists of vertical planks attached at floor- and roof-levels to horizontal beams. Although pattern books provided information about building practices during the nineteenth-century, most knowledge of construction was learned on the job, rather than from books.

Two related landscapes emerged as a result of the development of the lumber industry in California. Where redwood forests once stood there were only bald hillsides. As the trees disappeared from the landscape, cities and towns constructed primarily of wood appeared throughout the state. By the mid 1850s, when the commercial cores of these cities were rebuilt in brick, the residential districts that radiated from the city centers continued to be built of wood. Early images of these wooden cities, such as Carleton Watkin's photographs of San Francisco taken in the 1860s, show a powerfully stark landscape of plain wood houses stretching toward the horizon, over rolling, treeless hills.

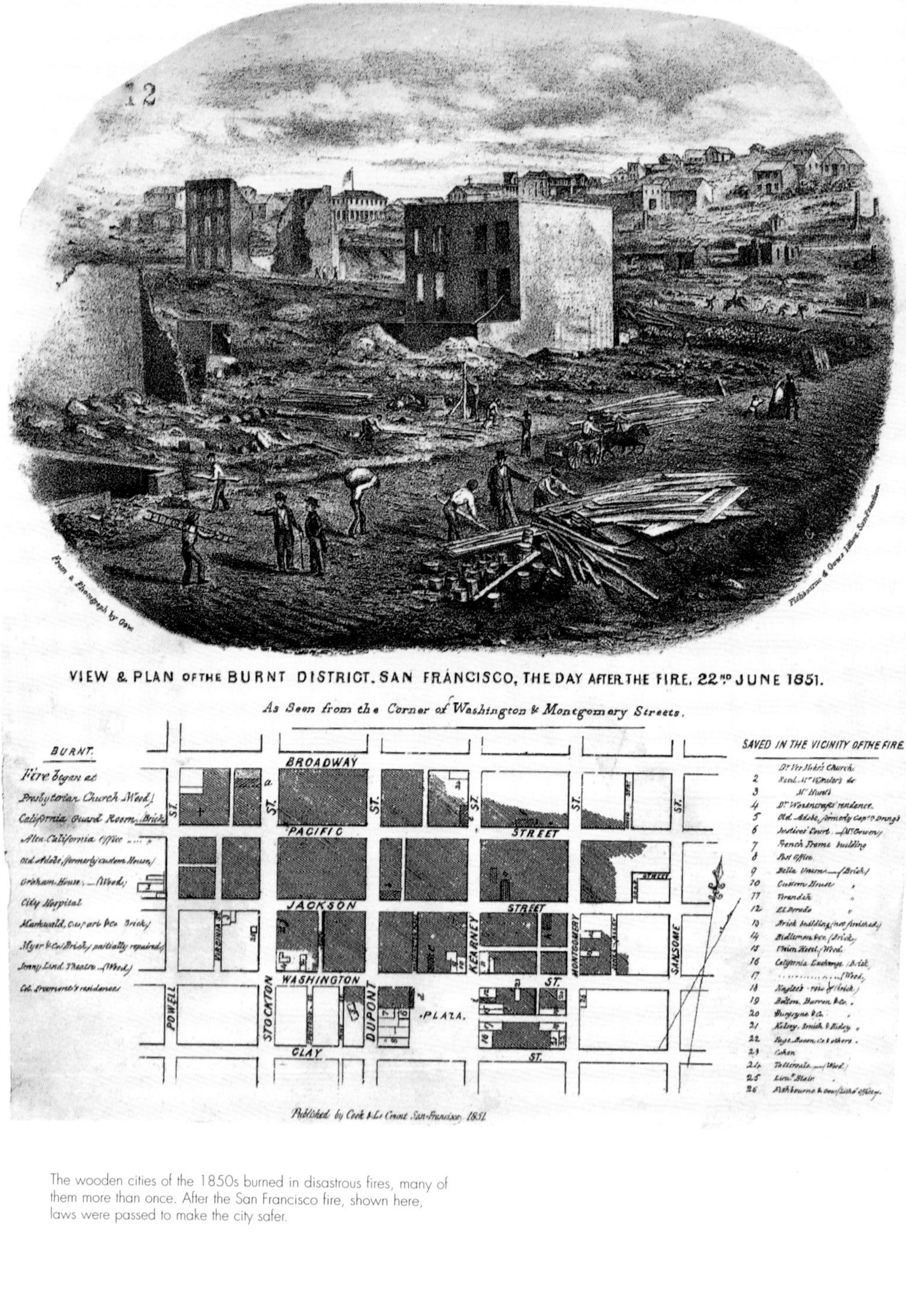

The wooden cities of the 1850s burned in disastrous fires, many of them more than once. After the San Francisco fire, shown here, laws were passed to make the city safer.

URBAN FIRES

Fires posed a widespread danger in American cities of the nineteenth century.[11] A number of conditions brought about by the industrial revolution contributed to the threat of urban fires: social disorder, the rapid growth of cities, and the introduction of new building materials and technologies. Almost every city and settlement built during the Gold Rush was badly damaged by fire between 1849 and the early 1850s. Columbia, Placerville, Marysville, Weaverville, Stockton, Sacramento, San Francisco, and other cities burned in large fires, some more than once. These fires frightened and discouraged the young population and contributed to the instability of life during the period.

The frontier cities and settlements of California were textbook cases of hazardous conditions: rows of wooden buildings, linked together by wooden sidewalks; poorly-constructed fireplaces and chimneys; oil and candle lighting with open flames; dangerous adjacencies of uses, such as the storage of gun powder and lighting oil within the same block as residences and saloons; and inadequate water and fire-fighting systems. The population of the Gold Rush consisted overwhelmingly of young males. These men were adventurers who were far from the stabilizing influences of home and often either excited by high expectations or discouraged by hardships. The early observations of public officials and medical doctors document a high incidence of drunkenness, emotional instability, and insanity among this population, which enhanced the likelihood that accidents, confusion, and rage would turn a spilled lamp into a large fire.[12]

Cities were established in California without any legal framework for building. After numerous fires, new municipal governments followed the example of other American cities in enacting laws to make buildings more fire-resistant. These laws required that hazardous materials and activities be moved outside of congested areas. They defined central areas of the city, called fire limits, within which buildings were required to be built in fireproof materials. They also required that new buildings incorporate fire safety features such as iron shutters placed over window and door openings and brick walls with parapets.

The first building laws in San Francisco, passed in 1850, were categorized within the city ordinances as fire laws. During the nineteenth century, the fire laws were frequently amended with new provisions for building. In addition, laws regulating construction were passed in other categories of the city's

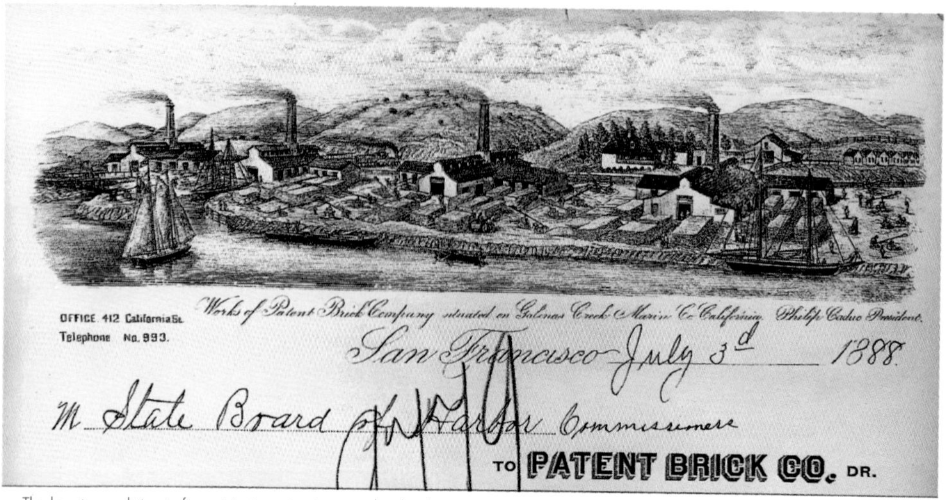
Works of Patent Brick Company situated on Galenas Creek Marin Co. California. Philip Caduc President.

OFFICE 412 California St.
Telephone No. 993.

San Francisco July 3d 1888.

M State Board of Harbor Commissioners

TO PATENT BRICK CO. DR.

The long-term solution to fire-resistant construction was the development of building industries in California. Factories such as the Patent Brick Company, featured on this billhead, shipped fireproof materials to all parts of California using the state's waterways.

ordinances. The result, by 1866, was a chaotic situation in which building laws were so scattered throughout the city's ordinances that builders could not easily use them and city authorities could not, or would not, enforce them.

Rising insurance rates also created pressure to construct fire-resistant buildings. Fire insurance companies sold their services by arguing that fire-resistant buildings, although more expensive to build than flammable buildings, were cheaper in the long run because insurance rates were significantly lower. These companies explained that, despite the use of the term "fireproof" in building laws and elsewhere, there was no such thing as fireproof construction; there were only degrees of fire-resistant construction, with insurance rates for each building set according to the character of its construction.

The first efforts to build fire-resistant buildings were hindered by the shortage of local fireproof building materials. As a temporary measure, in the 1850s and 1860s bricks, stone, cement, and iron were brought to California by ship. The long-term solution to fire-resistant construction would be the development of local sources of fireproof building materials.

Bricks were first produced along the Sacramento River and on the edge of San Francisco Bay, where deposits of clay were abundant. Fired bricks were loaded onto barges and easily transported to the waterfronts of San Francisco and other port cities. Because most early California brick was of poor quality, brick buildings were typically coated in stucco, both to improve their appearance and to preserve the brick. Eventually, good sources of clay and consistently high demand made it possible to produce a variety of high-quality brick. By the 1880s, a new process designed for the continuous firing of brick greatly increased the production capacity of brickyards in California.

Building in brick also depended on good mortar, composed of cement and water. California cement was first made from limestone. Although there were many producers of cement in California, the only one whose product could compete in quality with cements produced in the East during the period between 1860 and 1880 was located in Benicia.

Good building stone was observed in California during the Gold Rush. However, it was not economically feasible to transport stone from quarries located in remote areas to the cities and towns where it was needed until direct transportation links were established. When construction of the San Francisco City Hall began in 1871, the only quarry capable of supplying adequate stone was not a commercial operation, but the U.S. government's sandstone quarry located on Angel Island. With the construction of railroads in California in the 1870s, it became possible to obtain stone from all over California, and the commercial stone industry began to develop.

Building in brick or stone required skilled workers with a knowledge of masonry techniques. Stonemasons and bricklayers each possessed special expertise for structural and decorative masonry.[13] An important source for maintaining standards of quality and disseminating knowledge about bricklaying was the bricklayer's union, established in San Francisco in 1852.

Structural and decorative iron were manufactured in San Francisco for use in local buildings by the Sutter Foundry as early as 1853. After the Chicago fire of 1871 demonstrated that unprotected iron did not maintain its strength when exposed to a hot fire, California builders experimented with fireproofing techniques. During the 1870s, the architect David Farquharson surrounded iron columns with brick as a form of fire protection.

The various responses to the danger of large, urban fires transformed the wooden cities of the early Gold Rush within a few years into the relatively permanent cities that would characterize the 1850s and 1860s. The municipal laws of cities large and small described the boundaries of the fire limits. In small cities, the fire limits consisted of a corridor along the Main Street. In larger cities, like San Francisco and Sacramento, the fire limits were broad districts encompassing many blocks.

These new laws and practices did not end the risk of fire, but they did create safer cities that lasted much longer than their predecessors. Today, little survives of the first, quickly-constructed, wooden cities of the Gold Rush, but the fireproof cities that replaced them survive in many places. The main streets of Grass Valley, Nevada City, and Placerville, the Jackson Square district of San Francisco, and Old Town Sacramento represent the effectiveness of the first response to fire in the wooden cities of the Gold Rush.

The earthquakes of 1857, 1865, and 1868 did substantial damage to structures and frightened the public. This illustration shows damage that may have resulted from shoddy construction or from building on unstable ground.

EARTHQUAKES

The brick districts of California cities begun in the 1850s resisted fires far better than what preceded them. They were more vulnerable, however, to another danger that only came to be widely appreciated after a substantial amount of brick construction had been set in place—the earthquake.[14] Severe earthquakes that occurred during the Spanish-Mexican period, notably in 1808, 1812, 1836, and 1838, resulted from two principal faults—the Hayward and the San Andreas—which run northwest to southeast through the East Bay and San Francisco, respectively. Most newly-arrived Americans were only dimly aware of the danger posed by earthquakes. Among the few buildings from this period designed with earthquakes in mind were the Montgomery Block of 1853, with its diagonal braces of iron, and the Bank of North America of 1855, with an interior wood structure designed to absorb the movement of earthquakes.[15]

Earthquakes felt strongly in San Francisco in 1857, 1865, and 1868 did substantial damage to structures. Illustrations made at the time record preventable damage that may have resulted from shoddy construction such as poor foundations or weak mortar.[16] These illustrations also show the more serious problems inherent in brick or stone construction, such as collapsed walls.

With the earthquakes of 1865 and 1868, builders, engineers, architects, and entrepreneurs became conscious of the dangers of earthquakes and responded with a proliferation of construction techniques intended to resist their forces. Because the worst damage to result from earthquakes occurred in masonry buildings, several methods for improving masonry construction were proposed. Because masonry walls tended to break apart into many pieces, new systems were designed to tie individual bricks and stones into a unit that would remain in one piece. The most common proposals involved reinforcing masonry walls with grids of vertical bars and horizontal bands called bond iron. The earliest of these systems, used in a proposal for an iron-reinforced seawall by John Kelly, followed the 1865 earthquake. In 1869, William H. Foye patented a system of iron reinforcing for masonry walls that was used in many major buildings constructed before 1873, including the San Francisco City Hall, the United States Mint, the United States Subtreasury Building, and the United States Appraisers Store, all in San Francisco; the Lunatic and

After the 1868 earthquake, builders, engineers, architects, and entrepreneurs sought to develop better ways of building. The most widespread technique, known as bond iron, was used to reinforcing masonry walls at the U.S. Mint, pictured above.

Insane Asylum in Napa; and the Alameda County Courthouse. In 1870, David L. Emerson patented an earthquake-proof house that was similar in concept to Foye's bond iron system. John P. Gaynor's Palace Hotel, begun in 1870, utilized a different bond iron system, one that was so successful that it was difficult to demolish the building even after it was gutted by fire in 1906. David Farquharson's design for South Hall at the University of California in Berkeley, also begun in 1870, utilized still another bond iron system.

Like other nineteenth-century building technologies, systems for earthquake resistance were generally conceived not just as building improvements, but also as business ventures. Inventors patented their ideas in the hope that they would receive fees every time their invention was incorporated within a building. Patent holders frequently defended their right to protect the economic value of inventions in court.

The extent to which bond iron was incorporated in the masonry buildings of Northern California after 1868 is not known. Because it was rarely mentioned in newspaper articles or building records, and because it was hidden inside walls, it is difficult to know how often it was used. One possible indication of its use is the presence of an anchor—an iron plate usually round or star-shaped—on the outside of a wall. A common feature of Northern California brick buildings since the 1870s, the anchor may be part of a bond iron system, or it may simply be tied to an interior beam in a manner that helps the building to resist failures. The use of bond iron may have increased in 1901, when it was incorporated in the San Francisco building law for the first time. During the period of reconstruction that followed the 1906

earthquake and fire, the use of bond iron was required by law, but was not always used. It remained a feature of the building law until the 1930s.

The fear of earthquakes also resulted in the persistent use of wood in houses built for the wealthy, when houses for the upper class in other regions were often built in brick. Balloon-frame buildings generally did not fail in an earthquake because they moved as a unit. The most dramatic use of wood construction by the rich was the famous cluster of mansions on Nob Hill in San Francisco, built during the 1870s and 1880s for various railroad and mining millionaires. With the exception of the Flood Mansion, which was built with brick and bond iron, these mansions were constructed in wood, and all were destroyed by fire in 1906.

The response of architects and builders to the earthquakes of the 1860s affected the landscape in several ways. The fear of earthquakes prompted most people to build houses in wood. Although brick and stone were necessary for fire prevention and were required within the fire limits of cities, fears surrounding their performance in earthquakes discouraged their use and generated pressure to minimize the size of fire limits. California's built landscape was shaped in part by responses to the danger of earthquakes, but these responses often contradicted efforts to build for fire safety.

RANSOME'S
PATENTED CONCRETE APPARATUS
FOR MOULDING WALLS.

PATENTED MARCH 24TH 1885.

No. 314398.

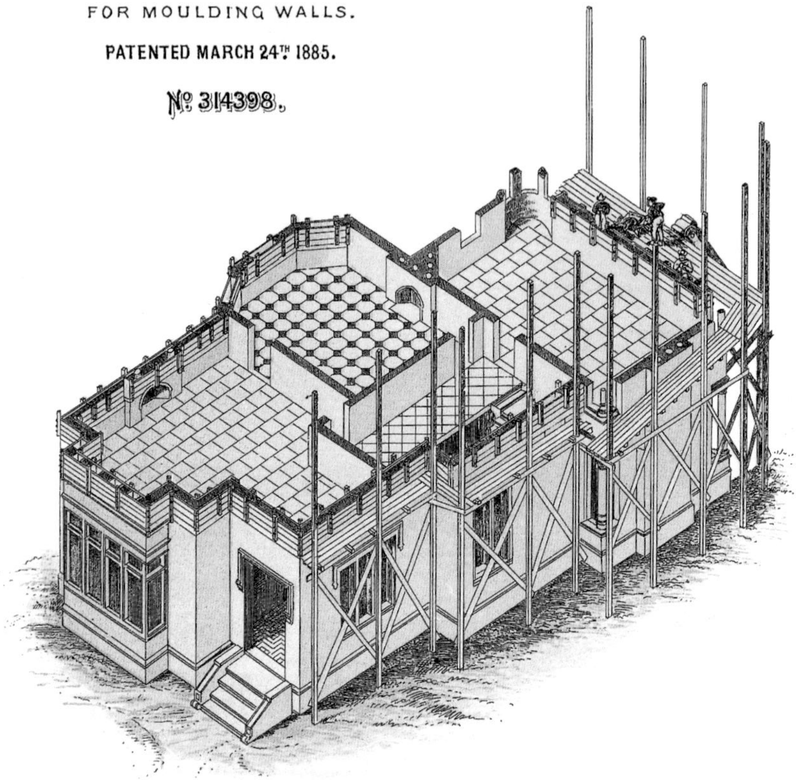

ISOMETRICAL VIEW OF BUILDING
IN COURSE OF ERECTION
With part of Scaffolding removed.

LITH. BRITTON & REY S.F.

+

Reinforced concrete was developed in California by Ernest L. Ransome. Ransome's patented methods for reinforcing concrete were first applied to industrial buildings in San Francisco and Alameda.

INDUSTRIAL EXPANSION AND MILITARY ASSISTANCE

The first large-scale structures built in Northern California were executed by industrial and military organizations.[17] These two types of organizations shared approaches to design and construction that differed from those of small-scale builders. Industrial organizations—often publicly-held corporations headquartered outside of California—invested large amounts of capital into large-scale building projects. During the second half of the nineteenth century, when the modern corporation emerged, organizations such as shipping companies, railroads, mining companies, warehousing and storage companies, mills and food processors, and manufacturers constituted both a new economic force and a major builder of structures. The efforts of the United States Army Corps of Engineers provided the infrastructure necessary to guarantee the success of many of these industrial enterprises.

The industrial and military facilities built in California during the nineteenth-century were designed by professionally trained engineers. These engineers had much in common with each other. Most were trained in a small number of universities, such as the University of California, Cornell University, and West Point. They were trained as military or civil engineers. They were not specialists, but generalists who could solve problems on a scientific and rational basis. Through professional societies and journals, they became familiar with a body of national and international engineering issues and standards. The Technical Society of the Pacific Coast served as a forum for the exchange of ideas by engineers from the U.S. Army, the University of California, Stanford University, and private practice. Through these connections, an important relationship developed during the nineteenth century between the military and those involved in the general industrial development of California.

The needs of industrial corporations generated a variety of building types; among these, the warehouse was the most common. Guided by extensive regulations contained in the building laws, warehouse design changed little over a period of nearly fifty years. From the Oriental Warehouse of the Pacific Mail Steamship Company of 1868, to the Nathan Dohrman Company Warehouse of 1912, both built in San Francisco, the best-built warehouses were of "slow-burning" mill construction. Mill construction consists of massive wood columns and beams, typically one foot square. Floors are laid directly on the beams to limit the amount of exposed surfaces and to minimize drafts. Exterior walls are constructed in brick and are self-supporting.[18] Heavy timbers proved to be superior to unprotected iron in surviving fires. Extensive warehouse districts of this "slow-burning" type arose in San Francisco along the north and south waterfront areas and in the area adjacent to the railroad yard along Townsend Street. Smaller districts existed in every port city in California, from small agricultural centers around San Francisco Bay such as Napa and Petaluma to larger regional distribution centers, such as Stockton and Sacramento.

The late nineteenth century was a period of experimentation in which groups of inventors and entrepreneurs played an important role in developing new materials and solving new problems. The inventors of bond iron systems comprise one such group. Several competing concrete contractors patented inventions for business purposes. Most notably, the inventor Ernest Ransome patented designs for square-twisted iron reinforcing bar in 1884 and an apparatus to build concrete walls in 1885. The first large-scale use of Ransome's methods occurred in reinforced-concrete industrial plants such as the 1884 Arctic Oil Works in San Francisco, which processed whale oil, and the 1888 Pacific Coast Borax Company in Alameda.[19]

The United States Army Corps of Engineers played an essential, if sometimes indirect, role in supporting the development of industrial corporations. San Francisco Bay was recognized as one of the great natural harbors of its time; but it, like other harbors around the world, required numerous improvements in order to accommodate new types of ships and new patterns of shipping associated with the industrial expansion of the late nineteenth century. The Corps of Engineers, responsible for all navigable waters within the United States, filled mud flats, built seawalls, dredged basins and channels, and blew up rocks that impeded navigation. At Mare Island, they built a large concrete dry dock. In the Oakland estuary, the Corps attempted to create a harbor by harnessing the tide.[20] The building of seawalls, dry docks, and other harbor works resulted in sophisticated solutions to many important engineering problems, especially those involving the use of concrete.

By 1900, military and industrial projects dominated the shoreline of San Francisco Bay and the navigable parts of the rivers that emptied into it. Military installations, navigational features, and military sources of cement, clay, and stone were located around the bay and on its islands. Warehouses and factories were given priority over other building types along city waterfronts, where railroads and ships met. Large industrial plants that engaged in dangerous activities, such as the manufacturing of explosives and lead, were spread out along isolated shorelines. Although their objectives were different, industrial organizations and the military had a similar, large-scale impact on the landscape of Northern California.

The U.S. Army Corps of Engineers filled mud flats, built seawalls, dredged basins and channels, and blew up rocks that impeded navigation. Here, the Corps builds Treasure Island, which was intended to become an airport.

ISSUED WITH S. F. NEWS LETTER CHRISTMAS NUMBER 1889

THE SAN FRANCISCO CHRONICLE'S NEW BUILDING. LITH BRITTON & REY S.F.

THOROUGHLY FIRE PROOF, LARGEST CLOCK IN THE WORLD, ENTIRELY LIGHTED BY ELECTRICITY.

Office buildings such as the Chronicle Building of 1888, built on an iron and steel frame and incorporating advanced electrical and mechanical feature, accommodated the expansion of business at the end of the nineteenth century.

BUSINESS EXPANSION

Parallel to the development of industry, nineteenth-century business expanded in California in tandem with the emergence of new types of organizations—expanded corporations, privately-held enterprises, and governments. Railroads and steamship companies, telegraph and telephone companies, newspapers, insurance companies, banks and building-and-loan societies, land companies, and governmental agencies grew in size and developed along new organizational lines.

These new or newly-expanded organizations were bureaucracies organized by department. They required a type of space not yet widely available in California; the response to this demand was the development of specialized office buildings. The office building was usually built by and named for a single company or individual, and provided office space for both the building's owner as well as additional tenants. Whereas during the Gold Rush office functions formed a small part of a company's larger operations and were accommodated in scattered locations—for example, a shipping company's office might be located on the company's pier—by the 1870s, the office functions of a variety of enterprises became concentrated within a specialized district. This centralized office district constituted a new kind of urban neighborhood and was given the name "downtown." The stacking of office floors in a single building and the clustering of office buildings within a specialized district reflected the desire of business owners to be located near competitors and those who provided business services.

The spatial requirements of office buildings were satisfied by means of new structural and mechanical technologies. Office buildings were typically three- to six-stories tall, with corridors providing access to rooms and suites of rooms on the upper floors. Located within a city's fire limits, they were constructed of fire-resistant materials such as brick, cast-iron, or heavy timbers and included brick or decorative cast-iron facades. Fire-resistant construction was considered highly desirable by the public, and some building owners—especially banks, building-and-loans, and insurance companies—advertised the fire-inhibiting features of their buildings.

In 1888, construction began on the iron- and steel-frame Chronicle Building in San Francisco, the first of a new and much taller type of office

Steel frame buildings required new fire-proofing technologies. In this illustration, hollow clay tile blocks clad steel members and provide fireproof floor structures.

building—the skyscraper. The Chronicle Building was followed in 1889 by the ten-story Mills Building and by many other skyscrapers constructed between 1898 and 1906. By 1906, the urban character of San Francisco was transformed: the features that rose above most of the buildings before 1890—the masts of ships, church steeples, factory stacks, the Selby Shot Tower, and the Palace Hotel—were dwarfed by tall office buildings that dominated the new skyline. Tall office buildings were made possible both by the use of structural skeletons of iron and steel and by the availability of electricity and modern mechanical systems. The iron and steel frame is capable of supporting the very heavy loads of a multi-story building without sacrificing space to thick, load-bearing walls. During this period, the columns and beams of the iron- and steel-frame building were connected with rivets and were designed to resist wind loads. The most prominent exception was the 1898 Call Building in San Francisco, with diagonal steel braces designed to resist earthquake stresses as well. As a means of protecting the building from fire, the frame was enclosed by brick exterior walls; floors were constructed of concrete or hollow tile; and individual steel members were clad in brick, hollow tile, plaster, or concrete.

Office buildings could not have functioned without the latest mechanical devices and lighting systems: running water, flush toilets, gas or electric lights, steam or hot-water heating, elevators, and telephones. These internal systems, together with external communications and transportation systems (such as streetcars), depended on the creation of a larger infrastructure. Plumbing, mechanical, electrical, communications, and transportation advances were developed, initially, through the enormous investment of private capital. In at least two areas, California cities were leaders in developing this infrastructure. In 1879, the first commercial electricity in the United States was produced near Third and Market streets in San Francisco and was made available for several blocks around. In 1883, Ernest Ransome built sidewalk vaults in Stockton, the first reinforced-concrete structures in the United States. The sidewalk vault is a segment of the sidewalk that spans the extension of a building's basement space, located below street level. The vault is penetrated by chutes, elevators, and stairs, designed for commercial deliveries. It is lit by natural light through glass lenses embedded within the concrete pavement. The sidewalk vault was an important development that contributed to the more intensive use of space in dense commercial districts. During the 1880s and 1890s, San Francisco was considered to have the best and most extensive concrete sidewalks in the United States.[21]

The building laws that regulated all of these developments changed constantly. Architects, engineers, and insurance companies advocated changes that would enhance building safety. Economic and social interests sometimes opposed such enhancements. Businesses wanting to build cheaply, labor unions protecting the rights of their members, and advocates for the poor—who could not afford to build fireproof buildings—often fought the addition of safety improvements to the building laws. These struggles resulted in the constant changing of regulations in response to opposing pressures. In the 1870s and 1880s, forces advocating the relaxation of building laws succeeded in reducing the size of San Francisco's fire limits and in modifying the regulations to allow for the construction of wooden houses within the fire limits. During the same years, efforts to organize the building laws scattered throughout the city's ordinances into a single, usable building code gradually succeeded; by 1901 the building laws were consolidated and the fire limits were again expanded.

The expansion of cities in the late nineteenth century resulted in a new urban landscape characterized by a proliferation of new building types and by a concentration of larger, taller buildings.[22] The predominant new building type—the office building—had a specialized use and was clustered in newly specialized districts within the city. These developments, made possible by the introduction of the iron and steel frame in tall buildings and by the adoption of a myriad of electrical and mechanical technologies, resulted in the creation of an urban landscape that would characterize the entire twentieth century. This was the city of specialized districts, including the centralized downtown with its skyline of tall office buildings.

Using earth, wood, iron, steel, terra cotta, brick, and concrete, engineers devised conduits to move water for many purposes. The large size and wood-stave construction of this pipe suggest that it was intended for irrigation or urban water supply.

MINING, WATER, AGRICULTURE AND POWER

During the late nineteenth and early twentieth centuries, several related technologies transformed the landscape of California.[23] A wide range of efforts at economic development utilized the technologies of mining, hydraulic engineering, hydroelectric power, and transportation. These various undertakings were interwoven by the engineers who designed them and by the owners of power companies, electric railroads, mines, lumber companies, and real estate agencies who controlled them. These intangible relationships of the individuals involved produced tangible connections among the various developments. These undertakings created strong ties between urban and rural areas and altered the landscape. In the cities, these undertakings enabled industrial development and population growth; in the countryside, the physical manifestations of these technologies became conspicuous features of the landscape.

The three principal approaches to gold mining—placer mining, hydraulic mining, and quartz mining—proved to be rich sources of new technology. The deep shafts of quartz mines were buttressed by timber frames and served by electrical and mechanical systems such as elevators and ventilation systems. These underground structures were sometimes larger and technologically more complex than urban buildings of the period: the design of mines may have led engineers toward solutions for the tall, steel-frame office building.[24] The development of wooden flumes used in placer and hydraulic mining provided valuable experience that was later applied to the design and construction of aqueducts, hydroelectric power plants, urban water reservoirs, and irrigation and land reclamation systems.

At the time of the Gold Rush, the pattern of winter rains, spring runoff, and summer drought produced seasonal lakes within the Central Valley, some as large as one hundred miles long and twenty miles wide. The high basements characteristic of late nineteenth-century Sacramento houses recall the period of routine flooding in this region. Hydraulic mining, which deposited tons of sediment into the state's rivers, exacerbated winter flooding.

As a consequence of the climate and topography, most California farmland was either too wet or too dry for farming. However, following the passage of the Wright Act of 1887, large tracts of land located in dry areas of the Central Valley were irrigated.[25] The Flood Control Act of 1911 facilitated the formation

Complex systems for supplying electricity and water, such as the hydroelectric dam at Hetch Hetchy, linked urban and rural areas of California.

of reclamation districts in flood-prone areas located in the center of the valley. These districts were encircled by levees and drained of flood water by means of canals and pumps, creating dry agricultural lands; subsequently, water for irrigation was pumped back in.

The provision of water and of hydroelectric power were often paired objectives made possible by a dam. Hydroelectric power was delivered from its sources in the Sierra Nevada to users in urban areas. Power was delivered across long distances on high steel transmission towers. On its way to the cities, electric power was sold to farmers, transforming farming methods and rural life in general. The earliest hydroelectric power use in Northern California was generated for the streetcar lines in Sacramento; subsequently, urban streetcars and interurban rail lines became major users of hydroelectric power.[26]

By the 1920s, the natural landscape of Northern California had become domesticated. The flat, alluvial lands of the Central Valley, the inland valleys, and the areas around the bay were generally overlaid with the grid of the federal land survey,[27] outlined in local roads, electric distribution lines, and the canals of irrigation and reclamation systems. Cutting across the parcels of land created by these relatively small-scale improvements were the high-voltage, long-distance, electric transmission lines; interurban electric rail lines and railroads, accompanied by telegraph and telephone lines; urban aqueducts; and the earliest of state and federal motor vehicle highways. This infrastructure not only linked the rural and urban landscape, but simultaneously transformed them: the urban skylines of steel-frame office buildings and the new rural landscape of logged forests, grazed grasslands, and reclaimed farmland were products of the same developments.

Long-distance, high voltage, electrical transmission towers transformed the rural landscape of the 1920s.

Reconstruction of modern, steel-frame buildings began soon after the earthquake, while reconstruction of structures in other materials was delayed until the approval of a revised building law in July 1906.

1906 DISASTER AND RECONSTRUCTION

The earthquake of April 18, 1906, devastated a broad area of western California.[28] In San Francisco, the earthquake caused a firestorm that destroyed much of the city. San Franciscans feared that the city would never recover and that terrible economic losses would ensue. To maintain public confidence and encourage investment, journalists and public officials emphasized fire damage over earthquake damage, believing that fire damage could be prevented. Professional investigations were initiated to try to understand the separate effects of the earthquake and the fire on buildings, and to identify the safest types of construction. Less than three months after the earthquake—before conclusions had been drawn—a new building code was adopted in San Francisco and reconstruction began. Within three years' time, San Francisco was largely rebuilt.

The new building code took effect on July 5, 1906, after study and debate by a committee of building professionals and politicians. Advocates of the working class, who feared that the poor could not afford to rebuild if required to build in brick, resisted efforts to enlarge the fire limits. In addition, the bricklayers' union resisted efforts to introduce reinforced concrete into the building code as an alternative structural technology. Despite these difficulties, the fire limits were expanded and some safety improvements made in the revised building code.

A year after the earthquake and fire, reports of expert investigators from the U.S. Geological Survey and the Roebling Construction Company compared steel-frame construction to what had previously been considered fireproof construction. The substantial-looking brick banks and insurance companies of the 1880s were reduced to ruins because their cast-iron facades and interior columns failed, and because wooden floors and partitions burned in the fire. By contrast, steel-frame buildings generally performed well in the earthquake and had mixed success in surviving the fire, depending in large part on the material used for fireproofing steel members. When the steel was covered in concrete, fire damage was minimal; when the steel was surrounded by hollow tile blocks, which cracked during the earthquake, leaving the steel exposed, fire damage was severe. The exterior materials used in buildings of all periods also performed with varying degrees of success in the fire: stone chipped and crumbled when subjected to fire, while terra cotta and brick held up well.

The most far-reaching lesson from the 1906 disaster was that reinforced concrete resisted damage from both earthquake and fire. The most dramatic example of the strength of reinforced concrete was the Bekin's Van and Storage Company warehouse, which was under construction at the time of the earthquake at the corner of 13th and Mission streets in San Francisco. Amid blocks of rubble, the Bekin's warehouse withstood both the earthquake and the fire. Its exterior brick walls were badly damaged, but its two-story interior structure of reinforced concrete columns, beams, and floors remained unharmed.

Building codes and practices changed, not only in San Francisco, but elsewhere, so that within a year after the earthquake, *The Architect and Engineer* stated that more reinforced-concrete buildings had been built in California than in any other place.[29] Particularly elaborate earthquake-resisting construction methods were adopted for the steel-framed Royal Globe Insurance Company Building of 1907, built in San Francisco. There, a mesh of reinforcing wire was placed inside its brick curtain walls. Many other buildings were built to the highest standards of fire-resistant construction of the day, designated Class-A construction in the building code. One such building, the Metropolitan Life Building of 1909, located on Nob Hill, included every fire-resistant feature available; its structural system included concrete foundations, floors, and roof, and a steel frame fireproofed in concrete; the frame was enclosed by brick exterior walls clad in glazed terra cotta; the windows were made with steel frames and wire glass; and the interior was built of fire-resistant materials, with steel lath and plaster partitions and marble and plaster finishes.

In San Francisco, these new building practices added new elements to the urban landscape. With greater reliance on steel frame construction, the city became taller. Stone, used previously to clad the city's best buildings, was now replaced by terra cotta, which glistened in the sun and could be washed to look always new. Where downtown streets were once dominated by brick walls, now the streets included many concrete buildings faced in stucco, usually painted a light color. The new, post-earthquake city was lighter than the brick and stone city of the past.

Outside San Francisco, rows of bungalows were built quickly to accommodate the masses of people who moved, temporarily or permanently, outside the city—many of them frightened by the loss of a home in the disaster. In contrast to the wooden board or shingle siding typical of houses built before 1906, fire-resistant stucco cladding was applied to many of these new houses. Portions of San Mateo County and the East Bay were changed radically, both by the amount of new development and by its new look.

The earthquake and fire of 1906 resulted in changes, both in the way buildings looked and in the way they were built. Despite the lessons learned from the disaster, however, many new buildings were built using older methods. Many people may have remained ignorant of improved building methods, and many others simply could not afford to implement them. Flammable wood-frame houses—many clad in wood siding—were still built

Until the late 1920s, San Francisco building laws required that steel frame structures be designed to resist wind loads only. The steel frame was assumed to be adequate for resisting earthquakes.

on unstable brick foundations. Commercial buildings continued to be built of load-bearing brick masonry walls with wood-frame interiors. While the earthquake and fire had a lasting impact on insurance and structural engineering practices, it produced an inconsistent response from architects and builders.

Architectural League Exhibition at Los Angeles

THE ARCHITECT AND ENGINEER OF CALIFORNIA

PACIFIC COAST STATES

PANAMA-PACIFIC INTERNATIONAL EXPOSITION SAN FRANCISCO 1915 CALIFORNIA

J.J.F.

$1.50 a Year
25c a Copy

PUBLISHED BY
The Architect and Engineer Co.
Entered at the Postoffice, San Francisco, as Second Class Matter

621 Monadnock Bldg.
San Francisco
Vol. XXVIII No. 2

MARCH, 1912

The professionalization of the building process was encouraged by journals like *The Architect and Engineer of California*. The cover of this issue highlights the role of technology, industry, and labor in modern construction.

PROFESSIONALISM, STANDARDIZATION AND REFINEMENT

At the beginning of the twentieth century, a variety of developments led to the growing professionalization of architecture and the building industry.[30] The prominent San Francisco engineer Walter T. Steilberg observed that during the first decade of the twentieth century there occurred a revolution in the way buildings were designed and built.[31] This revolution had to do not only with the way architects and builders practiced, but with new developments in technology. The components of this revolution were complex and had a pervasive impact on the landscape.

Beginning in 1901, architects in California could not practice without a license granted by the state. Partly in response to this, the San Francisco Architectural Club was established in the same year to help young draftsmen (and a few women) prepare for licensing while undergoing a traditional apprenticeship to a practicing architect. In 1903, informal courses in architecture at the University of California were replaced by a formal department of architecture. Together, these two systems of teaching served different classes of architects-in-training — those who could afford it went to the University, and those who could not worked during the day and studied architecture at night through the San Francisco Architectural Club or the Society of Beaux-Arts Architects.

Professional organizations and professional life flourished. The local chapter of the American Institute of Architects, founded in 1881, grew in membership and activity after 1901. A rival organization, the San Francisco Society of Architects, became active in the 1910s and 1920s. Both groups addressed professional concerns such as licensing rules, fees, and standards. At the same time, these groups functioned as social organizations. Meetings, banquets, outings, and parties (called "high jinks" and "low jinks") reinforced a sense of community among architects and a sense of separateness from those in other fields. In addition, leading architects became members of exclusive clubs, where business contacts were made and a sense of community was cultivated among the economic, social, and artistic elite of the region.

Scientific and technical knowledge increasingly served as the basis for building practices and standards. After 1903, research undertaken by the new Department of Architecture at the University of California began to influence local building practices and building laws. Several architectural journals, notably *The Architect and Engineer*, founded in 1905, provided sound technical information. The Office of the State Architect, established permanently in 1907, provided objective standards in building and design. In 1909, and every few years afterwards, refinements based on new technical knowledge were made to San Francisco's building laws, their incorporation reflecting the diminishing influence of non-technical considerations. After the Long Beach earthquake of 1933, the recommendations of engineers resulted in statewide adjustments to the building code.

The process of constructing large buildings underwent major changes as the building industry became subject to pressure from well-organized groups such as trade organizations, labor unions, and insurance companies. Construction companies such as Mahony Brothers built larger, more complex buildings. In about 1910, the City of San Francisco began hiring general contractors to coordinate large-scale projects. General contractors abandoned the prevailing practice of hiring day labor. In response to changing labor conditions and to the dangers of modern construction, the struggle to organize labor intensified.[32] By the time of the construction boom of the 1920s, most large buildings were built by highly-organized construction companies using union labor.

Engineers refined the structural features of buildings—the design of foundations, the design and placement of rebar, the way members of steel frames were attached and braced, and the way materials were fireproofed. The design of steel-frame structures changed little since the well-regarded Call Building of 1898, with its diagonal bracing, and the Humboldt Bank of 1907, with its deep spandrel girders. By 1906, the San Francisco building code provided standards for bracing tall buildings against wind; not until after 1927 did the code require additional measures to brace tall buildings against the effects of earthquakes. During the 1920s and 1930s, these refinements were incorporated in tall steel-frame buildings in San Francisco, Oakland, Berkeley, San Jose, Stockton, Sacramento, and Marysville.

Architects experimented extensively with decorative materials between 1906 and the 1930s. The architectural terra-cotta industry grew tremendously in California after 1906. Terra-cotta blocks were manufactured both to replicate carved stone in appearance and to carry loads within a masonry wall. Although the prototype for each block was expensive and time consuming to make, individual terra cotta blocks were much cheaper to produce than the alternative—blocks of carved stone. Following manufacturing innovations in the 1930s, most terra cotta was produced not as a load-bearing element, but as a veneer, no longer designed to replicate stone.[33]

Architectural sheet metal also decorated building facades and rooflines. Perhaps the most extravagant use of architectural sheet metal anywhere was installed between 1906 and the 1920s in San Francisco's Chinatown, where it took the form of cornices, pilasters, balconies, and pagoda roofs. Sheet metal was also produced as ductwork for heating, ventilating, and air conditioning systems.

Architectural education included both technical training and social events. Here the two are combined as the San Francisco Architectural Club tours McGilvray Quarry in Madera County.

Outdoor electric lights produced another kind of decorative effect. The Eastern Outfitting Company Building, located on Market Street, and several other buildings constructed between 1908 and 1909, were lit by incandescent bulbs placed along the skeletal frames of their facades. William D'Arcy Ryan, who described himself as the world's first illuminating engineer, was responsible for the innovative flood lighting of buildings at the 1915 Panama-Pacific International Exposition. In 1916, he created the Path-of-Gold streetlights along Market Street and the Triangle District streetlights in what is now San Francisco's retail district. The result, according to *The Journal of Electricity*, was that San Francisco had "the best lighted business district of any city in the world."[34] During the 1920s and 1930s, movie theater signs and marquees appeared as oases of light. The outdoor lights of the early twentieth century created a new kind of night landscape.

The landscape produced by these developments consisted of larger buildings than ever before, often with steel or reinforced-concrete structural systems. The electrical and mechanical systems that were introduced in a few buildings during the nineteenth century became universal features of early twentieth century buildings. In San Francisco, the devastation of the 1906 earthquake and fire hastened the modernization of buildings. The ornamental elements of buildings were increasingly the products of mechanical technology; contrary to appearances, which suggested the hand work of artists and craftsmen, most decorative features were produced by the skilled operators of machines. The physical changes that took place in the early twentieth century landscape reflected important changes in professional practice, the education of architects, and the organization of construction and labor.

The San Francisco skyline of the 1920s was a product of steel-frame construction and glazed terra cotta exterior walls.

The plywood industry developed slowly until World War II, when shortages of conventional building materials led to the widespread adoption of many new materials and technologies.

WORLD WAR II

World War II had a powerful effect on building conditions, not only within the military, but also within civilian society.[35] The immense amount of building that took place, the use of existing and new materials, and the processes of design and construction all had an enormous impact on the landscape of California.

Wartime conditions exacerbated a nationwide housing shortage that had been growing throughout the 1930s. Materials that might have been used for non-military housing were diverted to defense industries and to the military itself. Shortages of building materials resulted in the development of new materials and in the growing use of previously developed materials such as plywood, concrete block, cement-asbestos board, laminated wood, and aluminum. Materials shortages also led to increased experimentation with prefabricated buildings and building components.

On military bases, whole communities were rapidly constructed according to standardized plans developed to use scarce materials economically and to be built by a largely unskilled labor force. During the period prior to Pearl Harbor, only a handful of military bases existed in Northern California; by 1944, there were dozens. The establishment of each base required the rapid construction of buildings for housing, administration, maintenance, and operations. Standardized plans were prepared by each branch of the military; the Army, for example, developed standardized plans as the war progressed, called successively the 700 Series, the 800 Series, and the Modified Theater-of-Operations Series. A large post with hundreds of buildings, such as Fort Ord in Seaside, was built in a matter of months by following these plans.

The efficiencies required by the military were sought in civilian building as well. Civilian construction was largely limited to industries related to the war effort and to providing housing for workers in those industries. Throughout the Bay Area, in Sausalito, Vallejo, Richmond, and San Francisco, vast housing projects went up quickly for war industry workers and served as testing grounds for new ideas about the mass-production of housing. For the 5,000-unit Chabot Terrace and Carquinez Heights developments near Vallejo, the process of construction employed "[t]wo assembly lines, one based on wall-board technology, the other on plywood."[36] This represented a radical

Standardized plans made possible the rapid construction of simple buildings such as these, designed to meet the enormous need for military structures during the war.

departure from pre-war methods.

Much of the war-related worker housing was built as temporary housing and has been demolished. Permanent construction was largely reserved for military operations buildings and essential industrial plants. Large districts at the west ends of Oakland and Alameda, for example, were built up with warehouses, hangars, and other steel and concrete structures. Similarly, the shipyards in Richmond, Sausalito, Oakland, Alameda, and San Francisco were built as permanent facilities. Ironically, with the relocation of the shipbuilding industry to other parts of the world during the late twentieth century, little survives of these California shipyards today.

Almost all of this war-related construction took place within less than three years, from early 1942 to mid-1944. The nearly instantaneous appearance of so much new construction, matched only by the Gold Rush and the period of reconstruction after 1906, has had a lasting impact on the landscape. Dozens of large new military bases, each characterized by rows of nearly-identical, cheap, wood buildings, constituted a pervasive presence in the state. Major industrial complexes, including the largest shipbuilding operations in the world, established a new scale for industry around San Francisco Bay, and vast housing projects that dwarfed anything previously known expanded the built-up areas of Bay Area cities in a brand new and highly conspicuous fashion.

AFTER 1945

The period after World War II has been frequently described in terms of the enormous housing boom of the 1950s and the construction of sprawling automobile suburbs. These developments were facilitated by lessons learned through the efficient production of housing during World War II. At the same time, other changes were set in place at this time that reflect a shifting emphasis in design and building, away from the traditional concern for the structure of buildings and toward a concern for their electrical and mechanical systems.

A simple measure of this change in emphasis can be seen in the number of drawings required to produce a single building. During World War II, a building with complex technical requirements, such as an airplane hangar, could be presented in one drawing. Twenty years later, during the Cold War, an airplane hangar required dozens of drawings, most of them delineating electrical and mechanical systems.

World War II was a watershed event in the history of structures and materials in California building. Old technologies were discarded and new ones adopted on a large scale when construction resumed after the war. With the exception of concrete-block construction, the masonry technology disappeared. More elements of ordinary buildings, such as roof trusses, were prefabricated. The most radical changes may have come from improved building systems, such as lighting and air-conditioning, which made possible districts of windowless warehouses and industrial plants.

The impact of World War II on the use of structures and materials was comparable to that of other major events in California history: the establishment of missions, the Gold Rush, the urban fires of the 1850s and of 1906, and the earthquakes of 1868, 1906, and 1933. The relationship of these events to the history of structures and materials demonstrates the interdependence of building technology and the world around it. In California, the history of building technology is part of the continual remaking of the landscape in ways that are often indistinguishable from those employed in other parts of the United States and are sometimes quite distinct.

Notes

[1]John Brinckerhoff Jackson, *Discovering the Vernacular Landscape* (New Haven: Yale University Press, 1984), 3-13.

[2]Most information about native California buildings is generated as a by-product of studies whose principal focus is something other than the physical landscape. Nevertheless, much information about the physical landscape exists in such studies as Robert F. Heizer, ed., *California*, vol. 8, *Handbook of North American Indians*, William C. Sturdevant, ed. (Washington, D.C.: Smithsonian Institution, 1978), and Alfred Louis Kroeber, *Handbook of the Indians of California* (Washington, D.C.: U.S. Government Printing Office, 1925). Information contained in this section is also drawn from conversations with Malcolm Margolin, author and publisher of *News From Native California*.

[3]Information on building practices in the Spanish-Mexican period is drawn largely from Mardith K. Schuetz-Miller, *Building and Builders in Hispanic California 1769-1850* (Tucson: Southwestern Mission Research Center, and Santa Barbara: A Santa Barbara Trust for Historic Preservation, Presidio Research Publication, 1994), which is one of the best published studies on any aspect of California building history. Elisabeth L. Egenhoff ed., *Fabricas: A Collection of Pictures and Statements on the Mineral Materials Used in Building in California Prior to 1850* (N.p.: California Department of Natural Resources, Division of Mines, assembled as a supplement to the *California Journal of Mines and Geology* for April 1952) is a compilation of contemporary observations of California before the Gold Rush.

[4]Three excellent sources on building in this period are Harold Kirker, *California's Architectural Frontier*, 3rd ed. (Salt Lake City: Gibbs M. Smith, Inc., 1986); Charles E. Peterson, "Prefabs in the California Gold Rush, 1849," *Journal of the Society of Architectural Historians (JSAH)* 24, no. 4 (December 1965), 318-324; and David Harris and Eric Sandweiss, *Eadweard Muybridge and The Photographic Panorama of San Francisco. 1850-1880* (Montreal: Canadian Center for Architecture, 1993).

[5]Kirker, *California's Architectural Frontier*, 23-43.

[6]Peterson, "Prefabs in the California Gold Rush, 1849," 318.

[7]Kirker, *California's Architectural Frontier*; Harris and Sandweiss, *Eadweard Muybridge and The Photographic Panorama of San Francisco 1850-1880*; and Anne Bloomfield, "The Real Estate Associates: A Land and Housing Developer of the 1870s in San Francisco," *JSAH* 27, no. 1 (March 1978), 13-33. Many carpenters' and builders' handbooks, such as International Textbook Co.'s *Masonry, Carpentry, Joinery* (1899)(Chicago: Chicago Review Press, 1980), provide good descriptions of brace-frame and balloon- or platform-frame construction.

[8]Gunther Barth, *Instant Cities, Urbanization and the Rise of San Francisco and Denver* (New York: Oxford University Press, 1975).

[9]By the mid-1850s, a row of balloon-frame houses was built in Mendocino. One measure of its early use is that the balloon frame appears to have been used infrequently, if at all, in Benicia until the 1860s, when the supply of nails and standard lumber became available. In the 1860s, the balloon frame increasingly replaced brace-frame construction as planing mills were established in San Francisco. Planing mills produced lumber in standard dimensions from the rough lumber produced by saw mills. Planing mills also produced doors and window sashes, important elements in the mechanization of wood building construction. See Robert

Bruegmann, *Benicia: Portrait of an Early California Town* (San Francisco: 101 Productions, 1980), 6, 139.

[10]Bloomfield, "The Real Estate Associates," 16, 28.

[11]Kirker, *California's Architectural Frontier*; Michael R. Corbett, "The Effect of the Earthquake and Fire of 1906 on the San Francisco Building Law," (Environmental Design Library, University of California, Berkeley, 1980), photocopy. I am indebted to Anne Bloomfield for information about brick and cast-iron construction.

[12]Stuart A. Brody, "Hospitalization of the Mentally Ill During California's Early Years, 1849-1853" (Sacramento: Department of Mental Hygiene, n.d.), photocopy; Alva S. Klotter, "California Mental Hospitals, A Historical Sketch," *Bulletin of the Medical Library Association* 45, no. 2 (April 1957), 159-163; and E.T. Wilkins, *Insanity and Insane Asylums* (State Commission in Lunacy, prepared for Governor Henry M. Haight, 2 December 1871).

[13]Some of the earliest brick buildings were laid in running bond without headers. The first building laws relating to brick construction specified common bond with seven stretcher courses for each header course. By 1866, the use of common bond with five stretcher courses for each header course was established by law. This remained the standard until the 1930s, when brick was no longer used as a structural material.

[14]Basin Research Associates, *Archeological Monitoring and Architectural Documentation, San Francisco Main Library Project, Site of the Former City Hall Completed in 1897*, 2 vols. (San Francisco: Basin Research Associates, prepared for O'Brien-Kreitzberg and Associates, 1994). Information on bond iron presented in this section comes largely from publications written by and conversations with Stephen Tobriner.

[15]Kirker, *California's Architectural Frontier*, 84.

[16]Builders became aware of the danger of building on landfill, but no restrictions were placed on building in such areas.

[17]Joseph H. Hagwood, *Engineers at the Golden Gate* (San Francisco: U.S. Army Corps of Engineers, 1982); Technical Society of the Pacific Coast, Transactions (1884-1896).

[18]A common variant of mill construction, called semi-mill construction, appeared by 1900. In this type of construction, joists are placed over the beams in order to make wider bays possible. Forest Products Laboratory, *Wood Handbook* (Washington: United States Department of Agriculture, Forest Service, 1935), 213-215. Mill construction is described in most building codes and in many builders' handbooks of the late nineteenth and early twentieth centuries, such as Frank E. Kidder, *The Architects' and Builders' Handbook*, 17th ed., Thomas Nolan, ed. (New York: John Wiley & Sons, 1921).

[19]Ernest L. Ransome and Alexis Saurbrey, *Reinforced Concrete Buildings* (New York: McGraw Hill, 1912); Fred H. Hackett, ed., *The Industries of San Francisco* (San Francisco: Payot, Upham and Co., 1884).

[20]To accomplish this, they conceived of a system in which water would be captured at high tide and released at low tide through a channel defined by a pair of stone 'training walls'. The speed of the water flowing between the walls was intended to deepen the channel. Hagwood, *Engineers at the Golden Gate*.

[21]Corbett, "The Effect of the Earthquake and Fire of 1906."

[22]Associated with the development of corporations and office buildings was a proliferation of building types serving the various needs of a wealthy, consumer society. Among these types were dry goods stores, department stores, hotels, theaters, art museums, churches, schools, and libraries. In addition, important advances were made in the design of buildings associated with scientific advancement, such as slaughterhouses, food processing establishments, restaurants, and hospitals.

[23]Robert Kelley, *Battling the Inland Sea* (Berkeley: University of California Press. 1989), provides a good background for understanding changes in the Central Valley.

[24]I am indebted to Gray Brechin for this idea.

[25]Frank Adams, *Irrigation Districts in California, Reports of the Division of Engineering and Irrigation* (N.p.: California Department of Public Works, 1929).

[26]Hydroelectric power for Sacramento was generated at Folsom in 1895. The earliest hydroelectric power in California was provided to San Bernardino in 1891. A.H. Markwart, "Power in California," *Journal of the Franklin Institute* 204, no. 2 (August 1927), 148-150.

[27]Much of the land settled during the Spanish-Mexican period was subdivided in irregular parcels, originating in land grants. Nevertheless, even in the nine counties around the bay, much of the land was surveyed using the federal system of townships and ranges.

[28]United States Department of the Interior, Geological Survey, *The San Francisco Earthquake and Fire of April 18, 1906, and Their Effects on Structures and Structural Materials* (1907)(San Francisco: San Francisco Historical Publishing Company, n.d.); Corbett "The Effect of the Earthquake and Fire of 1906"; A.L.A. Himmelwright, *The San Francisco Earthquake and Fire, a Brief History of the Disaster* (N.p.: Roebling Construction Co., 1906).

[29]Horatio F. Stoll, "Reinforced Concrete Construction in California," *Cement Age* 4, no. 5 (May 1907), 314.

[30]California Department of General Services, Office of Architecture and Construction, *History of Office of Architecture and Construction* (Sacramento: California State Archives, 1968), photocopy.

[31]Walter T. Steilberg, *Miscellaneous Interviews with Walter Steilberg*, interview by Leslie Freudenheim and Elizabeth Sussman (Berkeley: Regional Oral History Office, Bancroft Library, University of California, 1976), 2-3.

[32]Winfield Hogaboom, "Heroes of Whom We Never Hear," *San Francisco Chronicle*(16 February 1908). The struggle to organize labor would culminated in strikes in 1921 and the mid-1930s.

[33]Artificial stone was a decorative material that was less important then terra cotta, but was widely used. Artificial stone was a type of concrete mixed with minerals and pigments to create color and texture. First used in the nineteenth century for sidewalks, garden walls, and landscape furniture, artificial stone became a much more authentic imitator of the original, as demonstrated by Paul E. Denivelle at the 1915 Panama-Pacific International Exposition and in the grand interiors of the San Francisco Public Library of 1916. For the next several decades, artificial stone was commonly used both as exterior trim and as an interior finishing material.

[34]A.F. Dickerson, "Lighting of San Francisco's Triangle District," *Journal of Electricity* 42, no. 9 (1 May 1919), 414-415.

[35]Diane Shaw Wasch, et al., *World War II and the U.S. Army Mobilization Program: A History of 700 and 700 Series Cantonment Construction* (N.p.: United States Department of the Interior, National Park Service, Cultural Resources, HABS/HAER, 1993).

[36]Roger Montgomery, "Mass Producing Bay Area Architecture," in *Bay Area Houses* , Sally Woodbridge, ed. (New York: Oxford University Press), 238.

Author's Acknowledgments

This essay is drawn in large part from the work of others. In addition to those cited in the notes and bibliography, conversations with the following people provided me with insights and information: Russ Bevill, Bob Bruegmann, Jim Buckley, Robert Clark, Margaret Henry, Ward Hill, Gary Knecht, Bill Kostura, Bridget Maley, Betty Marvin, Woody Minor, Virgil Nieman, Daves Rossell, Jack Schafer, Sally Woodbridge, and Daniel Zamalloa. I am especially grateful to the following, who commented on an early draft: Gray Brechin, Anne Bloomfield, Mary Hardy, Waverly Lowell, Malcolm Margolin, and Stephen Tobriner. Marlene McCall and Ruby Hendryx at Creative Office Services did the word processing. Waverly Lowell and Emily Wolff selected the illustrations. I have drawn extensively from previous research on industrial buildings that was generously supported by Luis Belmonte.

Bibliography

Adams, Frank, *Irrigation Districts in California, Reports of the Division of Engineering and Irrigation* (N.p.: California Department of Public Works, 1929).

American Institute of Architects, *A Handbook of Architectural Practice* (Washington, D.C.: Press of the American Institute of Architects, Inc., 1920).

American Society of Civil Engineers, San Francisco Section, *Historic Civil Engineering Landmarks*, William A. Myers, ed. (San Francisco: The Pacific Gas and Electric Co., 1977).

American Technical Society, *Cyclopedia of Architecture, Carpentry, and Buildings*, 10 vols. (Chicago: American Technical Society, 1912).

Arnett, F.A., *Electric Elevators*, 2nd ed. (New York: McGraw Hill, 1935).

Babcock & Wilcox Company, *Steam, Its Generation and Use*, 36th ed. (New York: Babcock & Wilcox Company, 1923).

The Architect and Engineer (May 1905-April 1945).

Barth, Gunther, *Instant Cities, Urbanization and the Rise of San Francisco and Denver* (New York: Oxford University Press, 1975).

Basin Research Associates, *Archeological Monitoring and Architectural Documentation, San Francisco Main Library Project, Site of the Former City Hall Completed in 1897*, 2 vols. (San Francisco: Basin Research Associates, prepared for O'Brien-Kreitzberg and Associates, 1994).

Bloomfield, Anne, "The Real Estate Associates: A Land and Housing Developer of the 1870s in San Francisco," *Journal of the Society of Architectural Historians* 27, no. 1 (March 1978), 13-33.

——"David Farquharson, Pioneer California Architect," *California History* 59, no. 1 (Spring 1980), 16-33.

Bonnett, Wayne, *Victorian San Francisco: The 1895 Illustrated Directory* (Sausalito, CA: Windgate Press, 1996).

Bradley, Denise, and Michael R. Corbett, *Rural Historic Landscape Report for Reclamation District 1000* (Chico, CA: Dames & Moore, prepared for the U.S. Army Engineer District, Sacramento, Corps of Engineers, 1995).

Brechin, Gray, "Ashes to Ashes," *San Francisco Chronicle* (24 November 1991).

Brody, Stuart A., *Hospitalization of the Mentally Ill During California's Early Years, 1849-1853* (Sacramento: Department of Mental Hygiene, n.d.), photocopy.

Bruegmann, Robert, *Benicia: Portrait of an Early California Town* (San Francisco: 101 Productions, 1980).

California Department of General Services, Office of Architecture and Construction, *History of Office of Architecture and Construction* (Sacramento: California State Archives, 1968), photocopy.

"California Rustic," *California Architect and Building News* (May 1885), 76-77.

Condit, Carl W., *American Building: Materials and Techniques from the First Colonial Settlements to the Present* (Chicago: University of Chicago Press, 1968).

Corbett, Michael R., "The Effect of the Earthquake and Fire of 1906 on the San Francisco Building Law," (Environmental Design Library, University of California, Berkeley, 1980), photocopy.

Corbett, Michael R., "Rearranging the Environment: The Making of a California Landscape, 1870s to 1990s," in *Facing Eden: 100 Years of Landscape Art in the Bay Area*, Steven A. Nash, ed. (Berkeley: University of California Press, and San Francisco: The Fine Arts Museums of San Francisco, 1995), 3-29.

Corbett, Michael R., *Splendid Survivors: San Francisco's Downtown Architectural Heritage* (San Francisco: A California Living Book, prepared by Charles Hall Page & Associates, Inc. for the Foundation for San Francisco's Architectural Heritage, 1979).

Corbett, Michael R., "The Use of Reinforced Concrete in California and the United States Before 1906," (Environmental Design Library, University of California, Berkeley, 1980), photocopy.

"Cut and Wire Nails Compared," *Pacific Rural Press* 50, no. 6 (10 August 1895), 95.

David, A.C., "The New San Francisco: Architectural and Social Changes Wrought by Reconstruction," *Architectural Record* 31, no. 1 (1912), 1-26.

Derleth, Charles, Jr., *The Destructive Extent of the California Earthquake: Its Effect Upon Structures and Structural Materials within the Earthquake Belt* (San Francisco: A.M. Robertson, 1907).

Dickerson, A.F., "Lighting of San Francisco's Triangle District," *Journal of Electricity* 42, no. 9 (1 May 1919), 414-415.

Donaldson, Barry, and Bernard Nagengast, *Heat & Cold, Mastering the Great Indoors, A Selective History of Heating, Ventilation, Air-conditioning and Refrigeration from the Ancients to the 1930s* (Atlanta: American Society of Heating, Refrigerating and Air-conditioning Engineers, Inc., 1994).

Draper, Joan Elaine, *The San Francisco Civic Center: Architecture, Planning, and Politics* (Ph.D. diss., University of California, Berkeley, 22 May 1979).

Egenhoff, Elisabeth L., ed., *Fabricas: A Collection of Pictures and Statements on the Mineral Materials Used in Building in California Prior to 1850* ([Sacramento]: California Department of Natural Resources, Division of Mines, assembled as a supplement to the California Journal of Mines and Geology for April 1952).

Elliott, Cecil D., *Technics and Architecture: The Development of Materials and Systems for Building* (Cambridge: MIT Press, 1992).

Farquharson, David, "Architecture Adapted to Earthquakes," *Daily Alta California* (19 February 1869), 1.

"Fire Proof Buildings: Jules Borie's Patent Hollow Bricks," *Pacific Rural Press* 12, no. 5 (29 July 1876), 81.

Fitzpatrick, F.W., and Theodore L. Condron, *Fireproof Construction* (Chicago: American School of Correspondence, 1914).

Forest Products Laboratory, *Wood Handbook* (Washington: United States Department of Agriculture, Forest Service, 1935).

Gayle, Margot, David W. Look, and John G. Waite, *Metals in America's Historic Buildings* (Washington, D.C.: U.S. Department of the Interior, National Park Service, Cultural Resources Preservation Assistance, 1992).

Gebhard, David, Eric Sandweiss, and Robert Winter, *Architecture in San Francisco and Northern California*, rev. ed. (Salt Lake City: Gibbs M. Smith, Inc., 1985).

Gillmore, Q.A., "Report on Béton Agglomére; or, Coignet-Béton and the Materials of Which it is Made," *Professional Papers, Corps of Engineers, U.S. Army*, no. 19 (Washington, D.C.: Government Printing Office, 1871).

Hackett, Fred H., ed., *The Industries of San Francisco* (San Francisco: Payot, Upham and Co., 1884).

Hague, William E., "San Francisco's Experience with Contracts on Public Works," *The Architect and Engineer* 18 (April 1912), 94-96.

Hagwood, Joseph H., *Engineers at the Golden Gate* (San Francisco: U.S. Army Corps of Engineers, 1982).

Hanson, Warren D., *San Francisco Water and Power; A History of the Municipal Water Department and Hetch Hetchy System* (San Francisco: The City and County of San Francisco, 1994).

Harris, David, and Eric Sandweiss, *Eadweard Muybridge and The Photographic Panorama of San Francisco. 1850-1880* (Montreal: Canadian Center for Architecture, 1993).

Hay, Duncan, and Michael Corbett, *Historic Resources Assessment Report for the Pit 1 Hydroelectric Project, Shasta County, California* (San Francisco: Dames & Moore, prepared for Pacific Gas and Electric Co., 1992).

Heizer, Robert F., ed., *California*, vol. 8, *Handbook of North American Indians*, William C. Sturtevant, ed. (Washington, D.C.: Smithsonian Institution, 1978).

Henry, Margaret, "Bay Area Brick and Terra Cotta from 1849 to 1976" (Environmental Design Library, University of California, Berkeley, 1976), photocopy.

Himmelwright, A.L.A., *The San Francisco Earthquake and Fire, A Brief History of the Disaster* (N.p.: Roebling Construction Co., 1906).

Hogaboom, Winfield, "Heroes of Whom We Never Hear," *San Francisco Chronicle* (16 February 1908).

International Textbook Co., *Masonry, Carpentry, Joinery*, in *International Library of Technology* (1899)(Chicago: Chicago Review Press, 1980).

Jackson, John Brinckerhoff, *Discovering the Vernacular Landscape* (New Haven: Yale University Press, 1984).

Jandle, H. Ward, ed., *The Technology of Historic American Buildings* (Washington, D.C.: The Foundation for Preservation Technology, for the Association for Preservation Technology, 1983).

Jenkins, Olaf P., ed., "Geologic Guidebook of the San Francisco Bay Counties," *Bulletin* 154 (San Francisco: California Department of Natural Resources, Division of Mines, 1951).

Jensen, Robert, "Board and Batten Siding and the Balloon Frame: Their Incompatibility in the Nineteenth Century," *Journal of the Society of Architectural Historians* 30, no. 1 (March 1971), 40-50.

Jester, Thomas C., ed., *Twentieth Century Building Materials* (Washington, D.C.: National Park Service and n.p.: McGraw Hill, 1995).

Kelley, Robert, *Battling the Inland Sea* (Berkeley: University of California Press, 1989).

Kidder, Frank E., *The Architects' and Builders' Handbook*, 17th ed., Thomas Nolan, ed. (New York: John Wiley & Sons, 1921).

Kirker, Harold, *California's Architectural Frontier*, 3rd ed. (Salt Lake City: Gibbs M. Smith, Inc. 1986).

Klotter, Alva S., "California Mental Hospitals, A Historical Sketch," *Bulletin of the Medical Library Association* 45, no. 2 (April 1957), 159-163.

Kroeber, Alfred Louis, *Handbook of the Indians of California* (Washington, D.C.: U.S. Government Printing Office, 1925).

Kurutz, Gary F., *Architectural Terra Cotta of Gladding, McBean* (Sausalito, CA: Windgate Press, 1989).

Leonard, John B., "Specifications for Reinforced Concrete," *The Architect and Engineer* 1, no. 4 (August 1905), 53-59.

"Lincrusta Walton," *California Architect and Building News* 5, no. 2 (February 1884), 29.

Lowell, Waverly B., Mary Hardy, and Lynn A. Downey, *Architectural Records in the San Francisco Bay Area, a Guide to Research* (New York: Garland Publishing, 1988).

Lowndes, William B., *Brick, Stone, and Plaster*, in *International Library of Technology* (Scranton: International Textbook Co., 1926), 31F.

Maginnis, Owen B, *Bricklaying* (New York: Owen B. Maginnis, 1900).

Mahony Brothers, *Illustrations of Buildings Built or Building During the Past 20 Months by Mahony Brothers of San Francisco* (N.p.: F.L. Ebbets, 1908).

Maley, M. Bridget, "The Patents of Ernest L. Ransome," paper presented at the Spring meeting of the Northern California Chapter, Society of Architectural Historians, Sonoma, CA, May 1997.

Margolin, Malcolm, *The Ohlone Way* (Berkeley, CA: Heyday Books, 1978).

——Editor's Notes, *News from Native California* 10, no. 1 (Fall 1996), 2.

Markwart, A.H., "Power in California," *Journal of the Franklin Institute* 204, no. 2 (August 1927), 145-192.

Marsh, Charles F., and William Dunn, *Reinforced Concrete*, 3rd ed. (New York: D. Van Nostrand Company, 1906).

Montgomery, Roger, "Mass Producing Bay Area Architecture," in *Bay Area Houses*, Sally Woodbridge, ed., (New York: Oxford University Press, 1976), 229-264.

Morrow, L.C., ed., *Maintenance Engineering Handbook*, 1st ed. (New York: McGraw Hill, 1957).

National Board of Fire Underwriters, *Dwelling Houses: A Code of Suggestions for Construction and Fire Protection*, 2nd ed. (New York: National Board of Fire Underwriters, 1920).

Newsom, Samuel, and Joseph C. Newsom, *Picturesque California Homes: A Volume of Forty Plates, Plans, Details and Specifications of Houses* (1884), introduction by David Gebhard (Los Angeles: Hennessey & Ingalls, Inc., 1978).

Nishkian, L.H., "Design of Tall Buildings for Resistance to Earthquake Stresses," *The Architect and Engineer* 88, no. 3 (March 1927), 73-83.

[Pelton, John Cotter], *Cheap Dwellings* (San Francisco: The San Francisco Bulletin Company, 1882).

Peterson, Charles E., "Prefabs in the California Gold Rush, 1849," *Journal of the Society of Architectural Historians* 24, no. 4 (December 1965), 318-324.

Pisani, Donald J., *From the Family Farm to Agribusiness, The Irrigation Crusade in California and the West 1850-1931* (Berkeley: University of California Press, 1984).

Powell, Edith Hopps, *San Francisco's Heritage in Art Glass* (Seattle: Superior Publishing Co., 1976).

Ransome, Ernest L., and Alexis Saurbrey, *Reinforced Concrete Buildings* (New York: McGraw Hill, 1912).

Robinson, W.W., *Land in California* (Berkeley: University of California Press, 1948).

Schuetz-Miller, Mardith K., *Building and Builders in Hispanic California 1769-1850* (Tucson: Southwestern Mission Research Center, and Santa Barbara: A Santa Barbara Trust for Historic Preservation, Presidio Research Publication, 1994).

Scott, Mel, *The San Francisco Bay Area, a Metropolis in Perspective* (Berkeley: University of California Press, 1959).

Simpson, Pamela H., "Cheap, Quick, and Easy: The Early History of Rockfaced Concrete Block Building," in *Perspectives in Vernacular Architecture* III, Thomas Carter and Bernard L. Herman, eds., (Columbia, MO: University of Missouri Press, 1989).

Steilberg, Walter T., *Miscellaneous Interviews with Walter Steilberg*, interview by Leslie Freudenheim and Elizabeth Sussman (Berkeley: Regional Oral History Office, Bancroft Library, University of California, 1976).

Stoll, Horatio F., "Reinforced Concrete Construction in California," *Cement Age* 4, no. 5 (May 1907), 314.

Sturgis, Russell, et al., *Sturgis' Illustrated Dictionary of Architecture and Building. An Unabridged Reprint of the 1901-2 Edition*, 3 vols. (New York: Dover Publications, 1989).

Svenson, Carl L., *A Handbook on Piping*, 2nd ed. (New York: D. Van Nostrand Co., 1922).

Technical Society of the Pacific Coast, Transactions (1884-1896).

Tobriner, Stephen, "Bond Iron and the Birth of Anti-Seismic Reinforced Masonry Construction in San Francisco," *The Masonry Society Journal* 5, no. 1 (January-June 1986), G12-G18.

United States Department of the Interior, Geological Survey, *The San Francisco Earthquake and Fire of April 18, 1906, and Their Effects on Structures and Structural Materials* (1907)(San Francisco: San Francisco Historical Publishing Company, n.d.).

Upton, Dell, "Pattern Books and Professionalism: Aspects of the Transformation of American Domestic Architecture, 1800-1860," *Winterthur Portfolio* 19, nos. 2-3 (Autumn 1984), 107-150.

Urquhart, Leonard Church, ed., *Civil Engineering Handbook*, 3rd ed. (New York: McGraw Hill, 1950).

Vance, James, *Geography and Urban Evolution in the San Francisco Bay Area* (Berkeley: Institute of Governmental Studies, University of California, 1964).

Wasch, Diane Shaw, et al., *World War II and the U.S. Army Mobilization Program: A History of 700 and 800 Series Cantonment Construction* (N.p.: United States Department of the Interior, National Park Service, Cultural Resources, HABS/HAER, 1993).

Watkins, C. Malcolm, *The White House of Half Moon Bay, James Johnston's Homestead, 1853* (Half Moon Bay, CA: The Johnston House Foundation, Inc., 1972).

Weber, Dickinson, *Downtown Street Views (c.1990) of Early Tall Buildings from the Bayside Cities and Tidewater Towns of Northern California, in Urban Architecture of the World: Downtown. The California Region*, no. 4 (Concord, CA: Sandscape Press, 1992).

——*Early Tall Buildings from the Agricultural Valley Towns of Central and Northern California: A Sentimental Sketchbook Collection, in Urban Architecture of the World: Downtown. The California Region*, no. 2 (Concord: Sandscape Press, 1988).

Whitney, C.W., "Ransome Construction in California," *The Architect and Engineer* 12, no. 3 (April 1908).

Wilkins, E.T., *Insanity and Insane Asylums* (The State Commission in Lunacy, prepared for Governor Henry M. Haight, 2 December 1871).

Wollenberg, Charles, *Golden Gate Metropolis, Perspectives on Bay Area History* (Berkeley: Institute of Governmental Studies, University of California, 1985).

Woodbridge, Sally B., *California Architecture: Historic American Buildings Survey* (San Francisco: Chronicle Books, 1988).

Woodward, George E., *Woodward's Country Homes*, 8th ed. (New York: G.E. & F.W. Woodward, 1866).

Ziebarth, Marilyn, "California's First Environmental Battle," *California History* 68, no. 4 (Fall 1984), 274-279.

List of Illustrations